Bibliographic information published by the German National Library:

The German National Library lists this publication in the National Bibliography; detailed bibliographic data are available on the Internet at http://dnb.dnb.de .

Imprint:

Copyright © 2014 GRIN Verlag, Open Publishing GmbH
Print and binding: Books on Demand GmbH, Norderstedt Germany
ISBN: 9783656870012

This book at GRIN:

http://www.grin.com/en/e-book/286684/wi-max-technology-a-survey

Hardik Modi, Chirag Umaretiya

Wi – Max Technology. A Survey

GRIN Publishing

GRIN - Your knowledge has value

Since its foundation in 1998, GRIN has specialized in publishing academic texts by students, college teachers and other academics as e-book and printed book. The website www.grin.com is an ideal platform for presenting term papers, final papers, scientific essays, dissertations and specialist books.

Visit us on the internet:

http://www.grin.com/

http://www.facebook.com/grincom

http://www.twitter.com/grin_com

Wi – Max Technology: A Survey

Chirag Umaretiya

Hardik Modi

stract— Worldwide Interoperability for Microwave cess (WiMAX) is a wireless communications technology. the global economy continues to expand, so does the nand for information. This information needs to be ntiful, diversified, instant, sprinkled and mobile, all at same time. These mounting demands have proven ﬁcult for existing network capacity, coverage and range o. And this reduces an organizations ability to evolve h changing business and workforce needs. Wireless works are best suited to meet these challenges because y are resilient, scalable, mobile and cost-effective. It ilitates the network with high-speed, high quality data, ce and video communication, even when users are on move. The rapid growth of WIFI in the home, erprise and public hotspot market has given users and vice providers alike a new glimpse into what wireless anectivity can deliver, through a single interface, in ltiple locations. WIFI realizes the vision of broadband anectivity within the LAN (Local Area Network) and easy way for new wireless technologies that can deliver same experience in the MAN (Metropolitan Area work) and in the WAN (Wide Area Network).

ywords– 5G; 4G; WWWW; LAN; MAN; WAN; Wifi; WiMAX

INTRODUCTION

rldwide Interoperability for Microwave Access (WIMAX) a wireless communications technology aim to supply eless data over long distances in a variety of ways as a imilar to cable and DSL (Digital Subscriber Line), from nt-to-point associations to occupied mobile cellular brand t to use. It is based on the IEEE 802.16 customary. [8] The ie WiMAX was produced by the WiMAX discussion, ich was twisted in June 2001 as an industry-led, not-for-fit association to encourage conformance and roperability of the customary. The ambition of this verable is to make available an impression of the ctionality and a explanation of the WiMAX network ctural design. [10] We study and evaluate the coexistence interoperability solutions stuck between WiMAX and r wireless access networks, such as WLAN (IEEE 802.11) nissing from 3G (B3G) networks. [13] We also appraise extraordinary facial appearance of the WiMAX nology, such as the superior treatment in Non Line Of it (NLOS) environments; in organize to examine the licability of well Known localization techniques. in clusion, we see the

sights the opportunity of increasing a new localization modus operandi that exploits the distinctiveness of WiMAX technology and the under misleading network transportation to distribute enhanced positioning accurateness. [11]

ARCHITECTURE

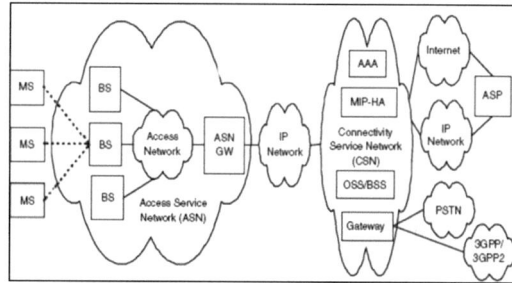

Fig – 1 WiMAX network Architecture [1]

MS – Mobile Station
BS – Base Station
ASN – Access Service Network
GW – Gate Way
OSS – Operation Support Subsystem
BSS – Base Station Subsystem
PSTN - Public Switched Telephone Network
3GPP – 3rd Generation Partnership Project

DIFFERENT TYPES OF DATA NETWORKS

Personal area network (PAN)

Personal area network is a normally wireless data network used for communication surrounded by data devices close to one person. The possibility of a PAN is then of the order of a few meters, generally understood to be less than 10 m, although WPAN technologies may have a greater reach. Bluetooth, UWB and Zigbee are the examples of WPAN. [6]

Local area network (LAN)

Local area network is a data network used for communication surrounded by data devices computer, telephones, printer and personal digital assistants. This network covers a moderately small area, like a home, an office or a small campus. The capacity of a LAN is of the order of

100 meters. The most currently used Lans are Ethernet and WiFi. [6]

Metropolitan area network (MAN)

Metropolitan area network is a data network that may perhaps cover up to more than a few kilometers, typically a large campus or a city. [6]

Wide area network (WAN)

Wide area network is ad data network covering a wide geographical region, as big as the planet. WANs are based on the correlation of LANs, allowing users in one location to communicate with users in other locations. Typically, a WAN consists of a number of interrelated switching nodes. These connections are made using leased lines and circuit – switched and packet – switched methods. The most at the moment used WAN is the internet network. Other examples are 3G and mobile WiMax networks, which are wireless WANs. The WANs repeatedly have much smaller data rates than LANs. [6]

RELEVANCE [20]

- Residential and SOHO high speed internet access.
- Small and medium business.
- WiFi hotspot backhaul.
- Frequent business traveler
- Living in multiple locations
- Fast web surfing and quick file download
- Real time video/audio streaming
- Interactive gaming
- VoIP (Voice over IP)
- Televise high definition TV
- Video on demand (VOD)

FEATURES [19]

- Long range
- Mobility
- Interfacing
- Accessibility
- Advance IP based Architecture
- Flexible channel Bandwidth
- QoS(Quality of Service) robust control
- Superior Performance
- Flexible
- Cost effectiveness
- Cost & CPE (Customer Premise Equipment) availability
- Smart antenna technology.
- Fractional frequency reuse
- Multicast and broadcast services.
- OFDM based physical layer
- Adaptive modulation and coding
- Support both FDD and TDD

PAYBACK OF WIMAX TECHNOLOGY [18]

- Lower development costs due to economies of scale.
- Reduced intimidation due to interoperability equipment Manufacturers.
- Stable supply of low cost technology and chips.
- Self-government to focus on spreading out of network fundamentals dependable with core competencies, while intentional that equipment will interoperate with third party products.
- Industrialized improvement efficiencies.
- Lower fabrication costs outstanding to economies of weighing machine Operators and examination Providers.
- Lower postulation hazard due to self-determination of choice in the middle of several vendors and solutions.
- Facility to adjust network to unmistakable applications by mixing and matching equipment from different vendors.
- Lower subscriber fees.
- Wider choice of terminals enabling cost presentation research.
- Portability of terminals when affecting locations/networks on or after WiMAX machinist "A" to machinist "B".
- Lower tune-up charge over point in time outstanding to asking price efficiencies in the liberation string.

WHY NEED WIMAX TECHNOLOGY?? [21 - 23]

- WiMAX technology brings true broadband connectivity to vertical applications.
- QoS allows operators to recommend prioritized access and SLAs.
- A single wireless crossing point supports voice and data services.
- WiMAX's end-to-end Internet Protocol (IP) core network facilitates combination with activity internal networks.

WIMAX VS WIFI

WiMAX operates on the identical wide-ranging ideology as WiFi. It sends data commencing single central processing unit to an additional via broadcasting signals. A central processing unit prepared with WiMAX would be given data beginning the WiMAX transmitting position, in all probability using encrypted information keys to avoid unconstitutional users from larceny admittance. [21 - 23] The greatest WiFi association can broadcast up to 54 megabits for each subsequent underneath the majority constructive location. WiMAX be supposed to be intelligent to switch up to 70 megabits for each following. Level some time ago that 70 megabits is separated up between more than a few dozen businesses or a small amount of hundred home users, it will make available at smallest amount the correspondent of cable-modem transport charge to everyone customer.

The prevalent difference isn't momentum it's remoteness. WiMAX outdistances WiFi by means of miles. WiFi's assortment is in relation to 100 feet (30 m). WiMAX will bedspread a radius of 30 miles (50 km) through wireless right to use. The augmented assortment is outstanding to the frequencies second-hand and the influence of the spreader. Of track, at that detachment, atmosphere, weather conditions and bulky buildings will take action to diminish the maximum assortment in a quantity of state of affairs, but the budding is in attendance to envelop enormous tracts of territory [21 - 23] WiMax is not premeditated to conflict with WiFi, but to coexist with it. WiMax treatment is deliberate in four-sided figure kilometers, at the same time as that of WiFi is deliberate in quadrangle meters.

WiMax provision in addition provides much superior services than WiFi, on condition that higher bandwidth and high data precautions by the make use of of superior encryption schemes. WiMax can also make available overhaul in equally Line Of Sight (LOS) and Non-Line of Sight (NLOS) locations, but the assortment will vary for that reason. WiMax will agree to the interpenetration for broadband overhaul stipulation of VoIP, video, and internet right of entry at the same time. WiMax can in addition work through accessible mobile networks. [21 - 23]

CHALLENGES TOWARDS WIMAX

- Most of the incidence spectrum anticipated for WiMAX accomplishment is already scattered by the government or reticent in favor of unambiguous purposes. As a outcome of this, WiMAX accomplishment has to be done on available advanced frequencies of the billed range for improved presentation.
- OFDMA is exceedingly disposed to time and frequency management errors. Insignificant frequency offset or little wait in timing will in addition outcome in a far above the ground bit blunder velocity subsequent to demodulation.
- Security Issues are established in WiMAX because of continuous communication between mobile station and base station to launch the connection. [15-16-17]

BENEFITS OF WIMAX [8-9-12]

- Guaranteed wide market acknowledgment of developed and Components.
- Lower production costs due to economies of amount.
- Reduced risk due to interoperability Equipment Manufacturers
- Unwavering contribute of stumpy cost mechanism and chips.
- Lack of restrictions to spotlight on expansion of set of connections fundamentals dependable with core competencies, while perceptive that apparatus will interoperate with third party foodstuffs.
- Engineering development efficiencies.
- Poorer invention expenditure outstanding to economies of weighing machine Operators and overhaul Providers.
- Poorer CAPEX – with poorer charge pedestal posting, purchaser location apparatus (CPE), and network exploitation expenditure.
- worse speculation jeopardy outstanding to lack of restrictions of selection among several vendors and solutions.

- Capability to modify network to explicit applications by incorporation and identical utensils from poles apart vendors.
- Enhanced machinist industry case with subordinate OPEX closing stages user.
- Lower subscriber fees.
- Wider preference of terminals enabling asking price routine investigation.
- Portability of terminals at what time moving locations/networks beginning WiMAX operative "A" to operative "B".
- Worse overhaul charge greater than point in time outstanding to cost efficiencies in the discharge chain.

WiMAX (802.16) PROTOCOL LAYER

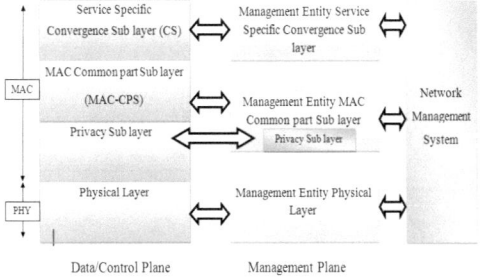

Fig – 2 WiMAX (802.16) Protocol Layers [2]

Physical Layer:
    The Physical Layer has been designed to support vast range of environments in the 10 – 60 GHz band. To follow a flexible spectrum usage, time division duplex and frequency division duplex configuration are supported along with the burst transmission format. The uplink is based on TDMA and DAMA (Demand Assign Multiple Access). The uplink is divided into multiple time slots for different users. [5]
    The number of time slots assigned to a particular user is determined by the MAC in the base station. The downlink channel is based on TDM with information about for the subscriber station attached to a stream of data and broadcast over a channel and received by all the station in the sector. [5]

OFDMA:
    Orthogonal Frequency division multiple access is a technique that divides the channel into multiple orthogonal sub channels. In this the input data stream is divided into several sub streams of a lower data rate and each sub stream is modulated and simultaneously transmitted on a separate sub channel. [5]

4

MAC Layer:

The MAC layer comprises of three subordinate layers. The overhaul unambiguous junction subordinate layer (CS) provides any information or mapping of exterior network data, compulsory from beginning to end the CS examine access point (SAP), into MAC SDUs (Service Data Units) received by MAC widespread Part subordinate layer (CPS) through MAC SAP. This includes classifying external network SDUs and associating the appropriate MAC examination flow identifier (SIFD) and connection identifier (CID). Multiple CS specification is provided for interfacing by way of an assortment of protocols. The MAC CPS provides the core MAC functionality of arrangement right of entry, bandwidth allowance, and relationship enterprise and relationship maintenance. It receives data commencing various CS through MAC SAP, classifies to particular MAC connection. The MAC also contains a separate security sub layer providing authentication. Secure key exchange and encryption. [5]

The Medium Access Layer is connection oriented. The fundamental premise of the MAC of the MAC layer architecture is the Quality of Service which is provided via the service flows. Unidirectional flow of packet is provided with a set of QoS parameters and applies to both downlink and uplink. The service specific convergence sub layer provide interface to higher layer protocols, classifies incoming etc. MAC common part sub layer includes the core MAC layer functions – scheduling, connection, maintenance, fragmentation and the QoS control. The privacy sub layer includes the functionalities like Encryption, Authentication and secure key exchange. [5]

COMPARISON OF 802.16 AND 802.20

Parameters	802.16e	802.20
Frequency Band	2-6 GHz	3.5 GHz
Channel Bandwidth	>5 MHz	<20 MHz
Transmission Rate	10 - 50 Mbps	>16 Mbps
Cell Radius	Up to 50 Kms	-
Mobility	High data rate fixed wireless user with adjunct mobility service	Fully mobile, high throughput data user
Mobile Speed	60 Kmph	Up to 250 Kmph
Services	Support of low latency data and real time voice services	Support of low latency data services
Roaming	Local regional mobility and roaming support	Global Mobility and roaming support
MAC/PHY	Extension to 802.16a MAC and PHY	New PHY and MAC optimized for packet data and adaptive antennas
Technology	Technology is optimized for and backward compatible with fixed station	Technology is optimized for full mobility

Table – 1 Comparison of 802.16 & 802.20 [4]

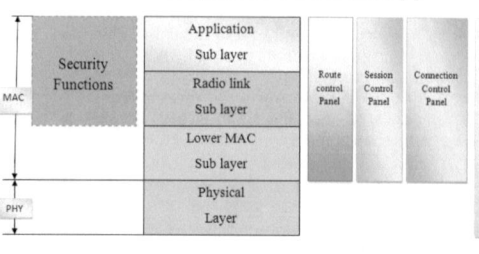

Fig – 3 Mobile Broadband layer Architecture [3]

CONCLUSION

The paper provided an analysis of the market issues surrounding WiMAX and identified challenges and opportunities for the development of WiMAX. It is clear that WiMAX has enjoyed some success in the wireless broadband landscape. Nonetheless, looking forward there are clear challenges ahead that are very important in behind WiMAX as a technology and for the development of WiMAX standards. The importance of OFDM has also been analyzed and this becomes an important feature that makes the difference between the 802.16 and 802.16a standard.

ACKNOWLEDGMENTS

We would like to thank the Charotar University of Science and technology for its constant support all the way through our work.

REFERENCES

[1] WiMAX network Architecture Available: - Wireless Communication by Madhav Ajwalia.

[2] WiMAX (802.16) Protocol Layer Available: - Wireless Communication by Madhav Ajwalia.

[3] Mobile Broadband layer Architecture Available: - Wireless Communication by Madhav Ajwalia.

[4] Comparison of 802.16 & 802.20Available:- Wireless Communication by Madhav Ajwalia.

[5] Physical layer, OFDMA and MAC layer Available: - Wireless Communication by Madhav Ajwalia.

[6] Different types of data networks Available: - Wireless Communication by Madhav Ajwalia.

[7] Shyy, D.J.; Ma, J.; Refaei, M.T., "WiMAX RF planner," *Testbeds and Research Infrastructures for the Development of Networks & Communities and Workshops, 2009. TridentCom 2009. 5th International Conference on* , vol., no., pp.1,3, 6-8 April 2009

[8] Ansari, A.; Dutta, S.; Tseytlin, M., "S-WiMAX: adaptation of IEEE 802.16e for mobile satellite services," *Communications Magazine, IEEE* , vol.47, no.6, pp.150,155, June 2009

[9] A socio-economic analysis of WiMAX," *Application of Information and Communication Technologies, 2009. AICT 2009. International Conference on* , vol., no., pp.1,6, 14-16 Oct. 2009

[10] Chien-Hung Yeh; Chi-Wai Chow; Yen-Liang Liu; Sz-Kai Wen; Shi-Yang Chen; Chorng-Ren Sheu; Min-Chien Tseng; Jiunn-Liang Lin; Dar-Zu Hsu; Sien Chi, "Theory and Technology for Standard WiMAX Over Fiber in High Speed Train Systems," *Lightwave Technology, Journal of* , vol.28, no.16, pp.2327,2336, Aug.15, 2010

[11] Martin, J.; Bo Li; Pressly, W.; Westall, J., "WiMAX performance at 4.9 GHz," *Aerospace Conference, 2010 IEEE* , vol., no., pp.1,8, 6-13 March 2010

[12] Pareit, Daan; Lannoo, Bart; Moerman, Ingrid; Demeester, Piet, "The History of WiMAX: A Complete Survey of the Evolution in Certification and Standardization for IEEE 802.16 and WiMAX," *Communications Surveys & Tutorials, IEEE* , vol.14, no.4, pp.1183,1211, Fourth Quarter 2012

[13] I-Hsuan Peng; Yen-Wen Chen; Chang-Wu Chen; Yu-Chin Huang, "Management of multiplexed ATM connection over WiMax network," *Network Operations and Management Symposium (APNOMS), 2011 13th Asia-Pacific* , vol., no., pp.1,5, 21-23 Sept. 2011

[14] Jha, R.K.; Dalal, U.D., "Location Based Radio Resource Allocation (LBRRA) in WIMAX and WLAN network," *Information and Communication Technologies (WICT), 2011 World Congress on* , vol., no., pp.399,406, 11-14 Dec. 2011

[15] Sedani, B.S.; Kotak, N.A.; Borisagar, K.R.; Kulkarni, G.R., "Implementation and Performance Analysis of Efficient Wireless Channels in WiMAX Using Image and Speech Transmission," *Communication Systems and Network Technologies (CSNT), 2012 International Conference on* , vol., no., pp.630,634, 11-13 May 2012

[16] Patidar, M.; Dubey, R.; Jain, N.K.; Kulpariya, S., "Performance analysis of WiMAX 802.16e physical layer model," *Wireless and Optical Communications Networks (WOCN), 2012 Ninth International Conference on*, vol., no., pp.1,4, 20-22 Sept. 2012

[17] Marasevic, J.; Janak, J.; Schulzrinne, H.; Zussman, G., "WiMAX in the Classroom: Designing a Cellular Networking Hands-On Lab," *Research and Educational Experiment Workshop (GREE), 2013 Second GENI* , vol., no., pp.104,110, 20-22 March 2013

[18] WiMAX: Features and Applications Available
http://www.google.co.in/url?sa=t&rct=j&q=&esrc=s&source=
web&cd=1&ved=0CBwQFjAA&url=http%3A%2F%2Fwww.
agir.ro%2Fbuletine%2F687.pdf&ei=m6ppVPKZJI-
eugSl5oHoBg&usg=AFQjCNEWn369RBjnt7OcAtMcnFtaJI
C6ow&bvm=bv.79142246,d.c2E

[19] IEEE 802.16: WiMAX Overview, WiMAX Architecture Available
http://www.ijcte.org/index.php?m=content&c=index&a=show
&catid=51&id=919

[20] WiMax technology and its applications Available
http://www.google.co.in/url?sa=t&rct=j&q=&esrc=s&source=
web&cd=1&ved=0CBwQFjAA&url=http%3A%2F%2Fwww.
ijera.

[21] Introduction to WiMAX Technology Available
http://www.google.co.in/url?sa=t&rct=j&q=&esrc=s&source=
web&cd=4&ved=0CC8QFjAD&url=http%3A%2F%2

[22] OFDMA WiMAX Physical Layer Available
http://www.google.co.in/url?sa=t&rct=j&q=&esrc=s&source=
web&cd=2&ved=0CCcQFjAB&url=http%3A%2F%2Fwww.s
pringer.

[23] WiMax Overview Available
https://www.cs.umd.edu/class/fall2009/cmsc417/Slides/WiM
AX-class.pdf

YOUR KNOWLEDGE HAS VALUE

- We will publish your bachelor's and
  master's thesis, essays and papers

- Your own eBook and book -
  sold worldwide in all relevant shops

- Earn money with each sale

Upload your text at www.GRIN.com
and publish for free